Mikoyan Gurevich MiG-21MF

1/72

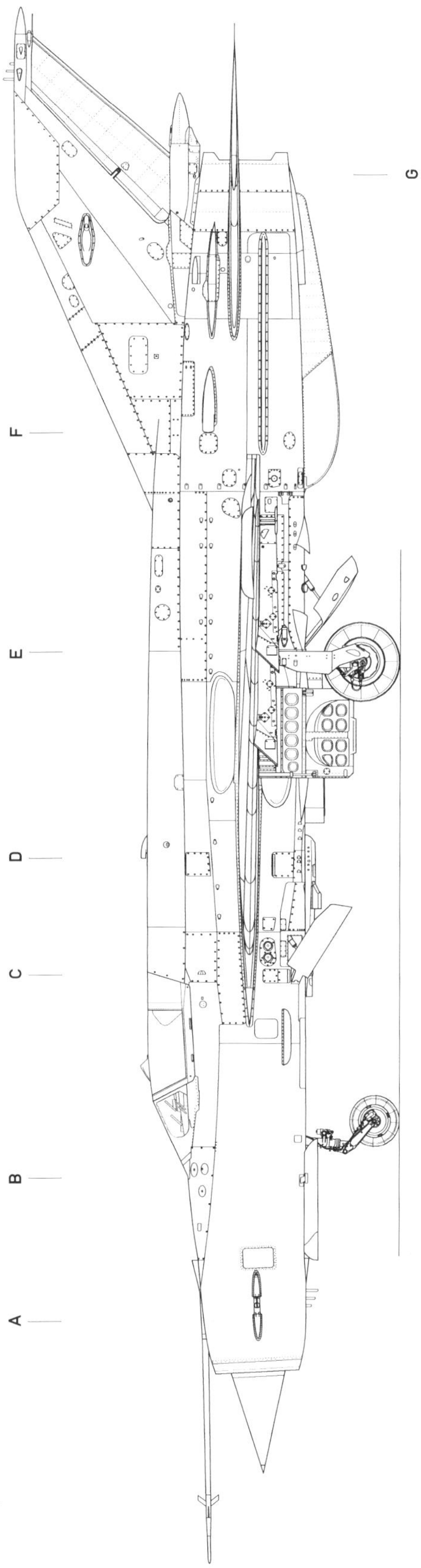

MiG-21MF.

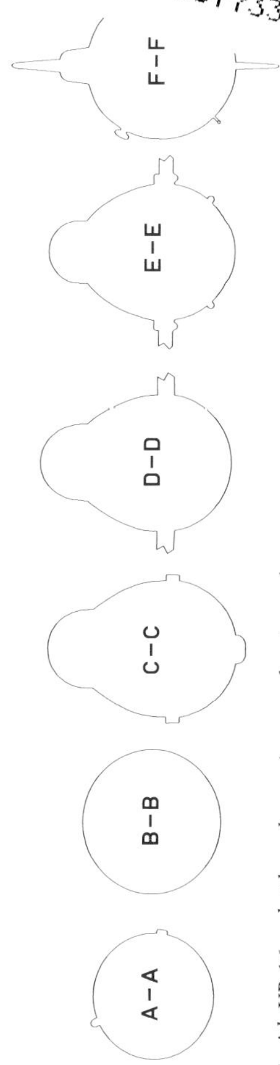

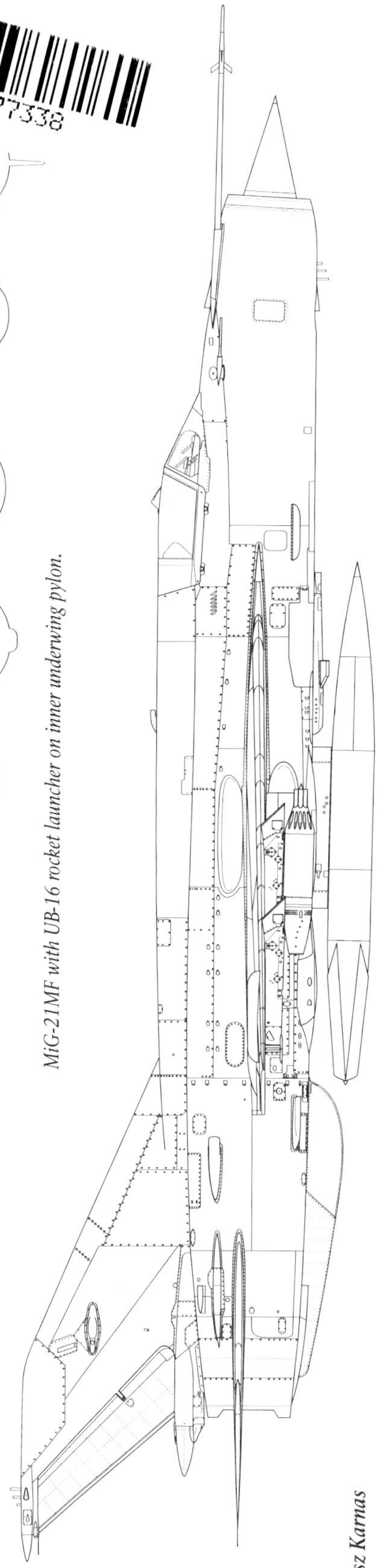

MiG-21MF with UB-16 rocket launcher on inner underwing pylon.

Dariusz Karnas

1/72

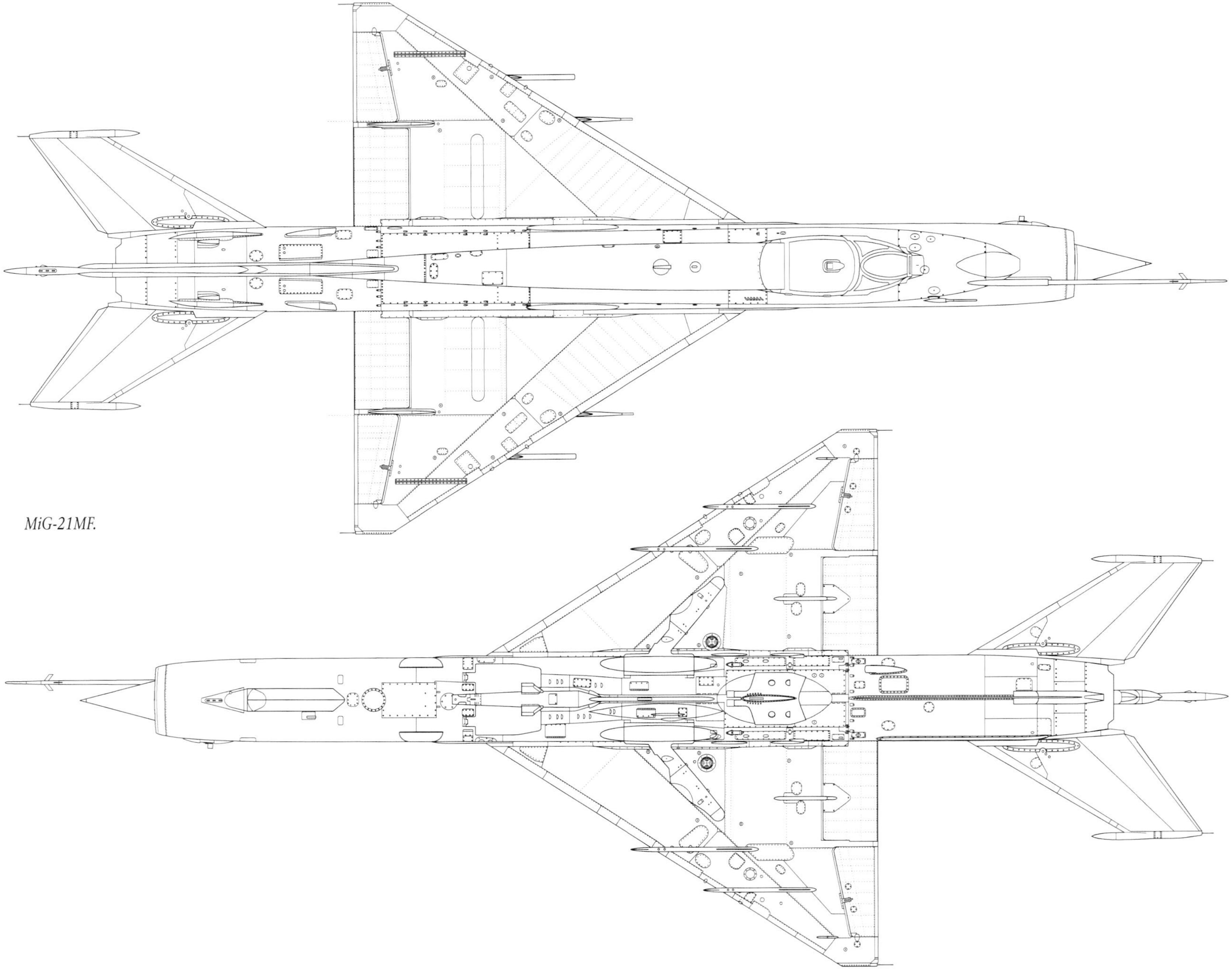

MiG-21MF.

Dariusz Karnas

1/72

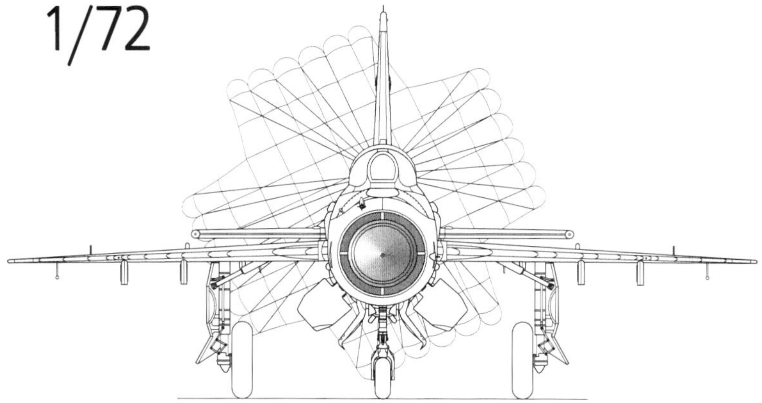

MiG-21MF.

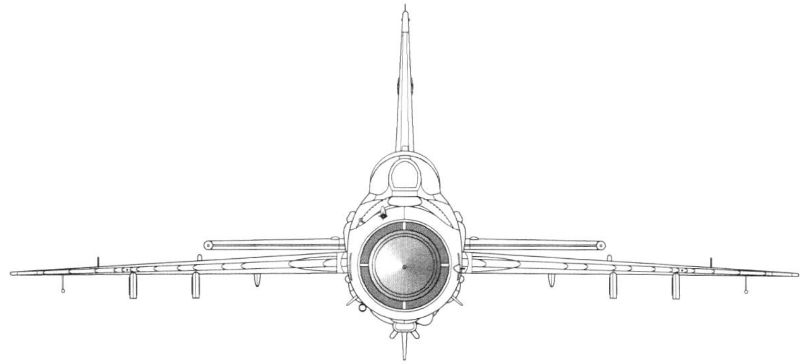

Armament and fuel tanks variants.

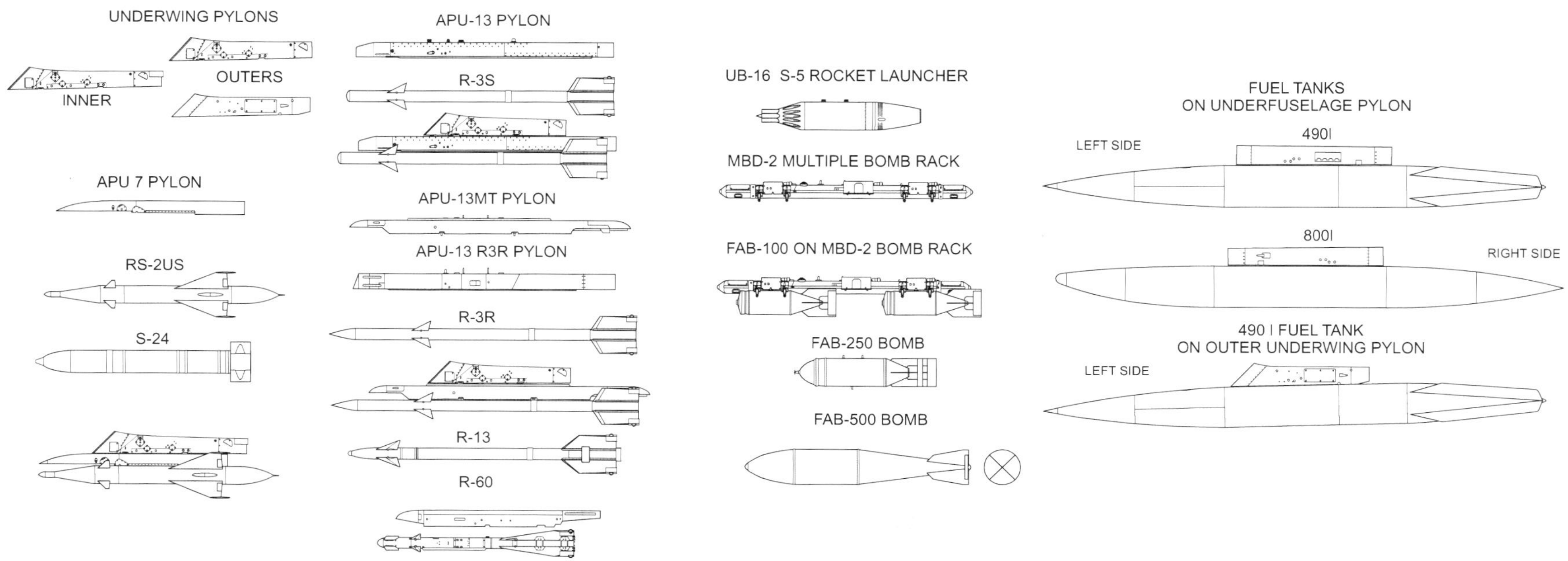

 Dariusz Karnas

MiG-21MF "9114" (s/n 96A9114) 41. PLM in Malbork (Pułk Lotnictwa Myśliwskiego – Fighter Aviation Regiment), delivered to Poland on 7 May 1974, decommissioned in 2003. (Wacław Hołyś)

MiG-21MF "8015" (s/n 96A8015), delivered to Poland in 11 September 1973, decommissioned in 2003. (Wacław Hołyś)

MiG-21MF "7906" (s/n 96A7906) 41. PLM in Malbork, delivered to Poland on 17 July 1973, decommissioned in 2003. (Dariusz Karnas)

MiG-21MF "9107" (s/n 96A9107) 10. elt in Łask (eskadra lotnictwa taktycznego – Tactical Squadron), delivered to Poland on 28 March 1974, decommissioned in 2003. (Dariusz Karnas)

MiG-21MF "6804" (s/n 96A6804) 10. elt in Łask, delivered to Poland in 7 December 1972, decommissioned in 2003. (Dariusz Karnas)

MiG-21MF "9011" (s/n 96A9011), 10. PLM in Łask, delivered to Poland on 24 February 1974, decommissioned in 2003. (Wacław Hołyś)

MiG-21MF "9114" (s/n 96A9114) 62. PLM in Poznań - Krzesiny, delivered to Poland on 28 March 1974, decommissioned in 2003. (Leszek Jóźwik)

MiG-21MF "6814" (s/n 96A6814), 10. PLM in Łask, delivered to Poland on 28 March 1973, decommissioned in 2003. (Marek Idzior)

Fuselage

Front of fuselage. Nose wheel doors and IFF antennas are visible. (Dariusz Karnas)

Mid part of fuselage. Nose wheel debris deflectors are visible. (Dariusz Karnas)

Rear fuselage details. (Dariusz Karnas)

Front air brakes (both sides) in open position. (Dariusz Karnas)

External 28,5 V DC and 115 V AC power connectors. Air brakes and GSch-23 cannon are also visible. (Dariusz Karnas)

Angle of attack sensor. (Dariusz Karnas)

Air data boom with Pitot tube and pitch and yaw vanes (sensors). Auxiliary dynamic pressure probe is also visible. (Dariusz Karnas archive)

Front fuselage section without engine. Rear air brake in open position. An actuator is disconnected. (Dariusz Karnas archive)

R-13-300 afterburning turbojet engine. (Dariusz Karnas)

Tail

Starboard all-moving horizontal stabilizer with anti-flutter weight. (Dariusz Karnas archive)

Vertical stabilizer. On the top it's IFF antenna, VHF/UHF R-802W dielectric fairing and MRP-56 system antenna. All moving horizontal stabilizer linkage fairing and brake parachute housing is visible. (Dariusz Karnas archive)

Split conical fairing parachute doors. (Dariusz Karnas archive)

Horizontal stabilizers and rear fuselage details. Exhaust nozzle fairing is visible. (Leszek Jóźwik)

Split conical fairing parachute doors in open position. (Dariusz Karnas)

Port all-moving horizontal tailplane with anti-flutter weight. (Dariusz Karnas)

Wing

Under side of the MiG-21MF wings. (Dariusz Karnas)

Starboard wing. Aileron hinge control linkage fairing and wing fence are visible. (Dariusz Karnas)

Port wing. The static discharger on wing tip. (Dariusz Karnas)

Cockpit & Canopy

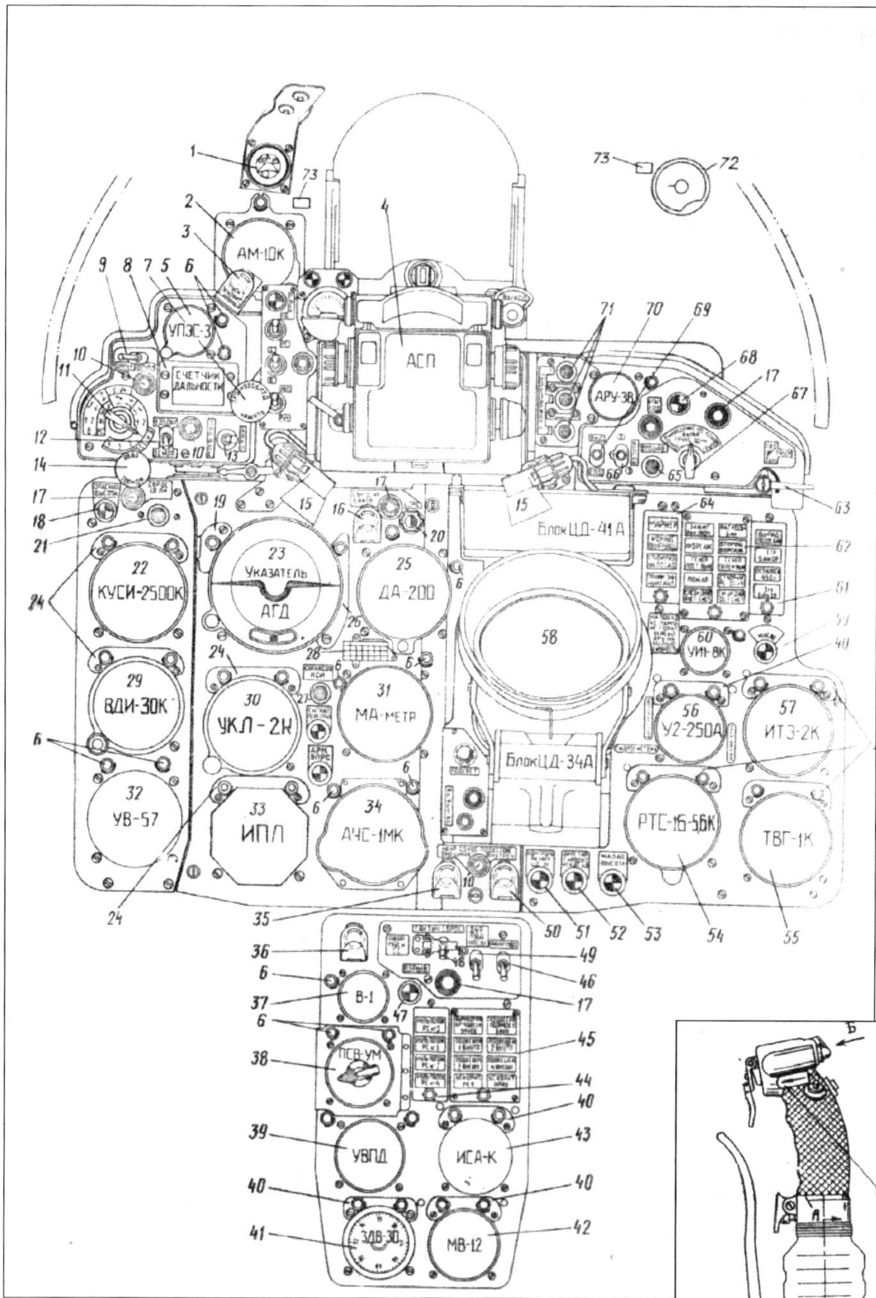

Instrument panel. (Technical Manual)

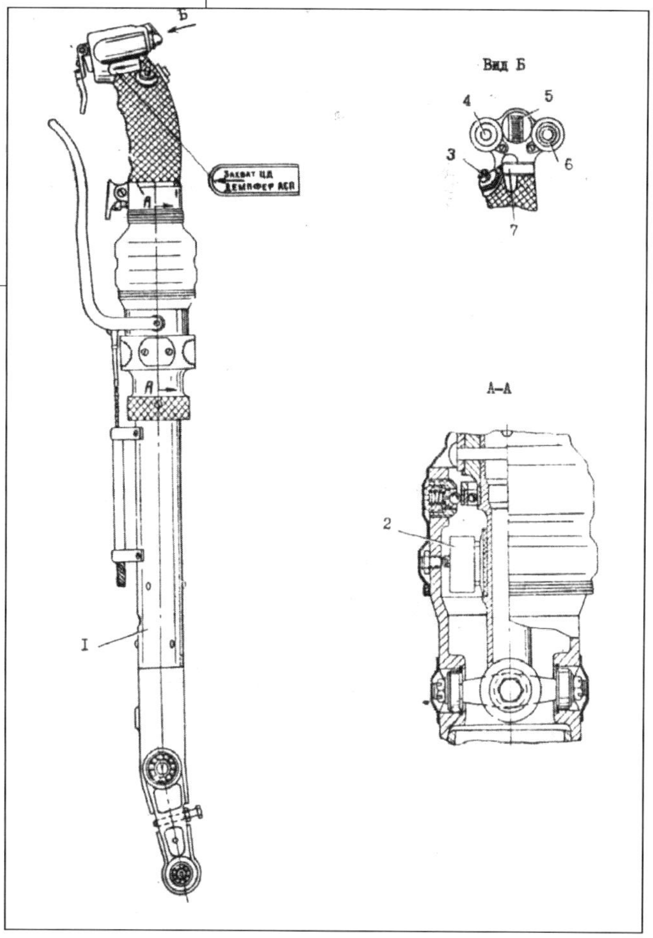

MiG-21MF controls, control column (right) and engine throttle lever (below). (Technical Manual)

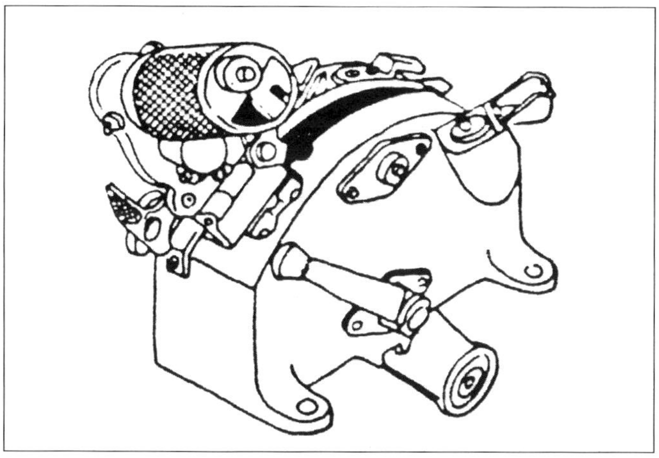

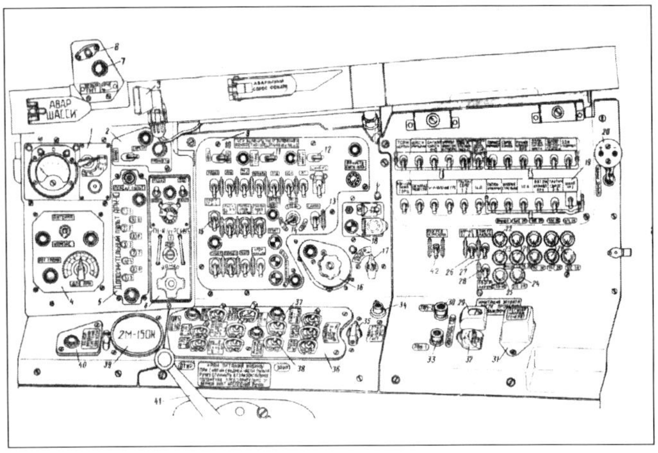

Starboard side of the cockpit. (Technical Manual)

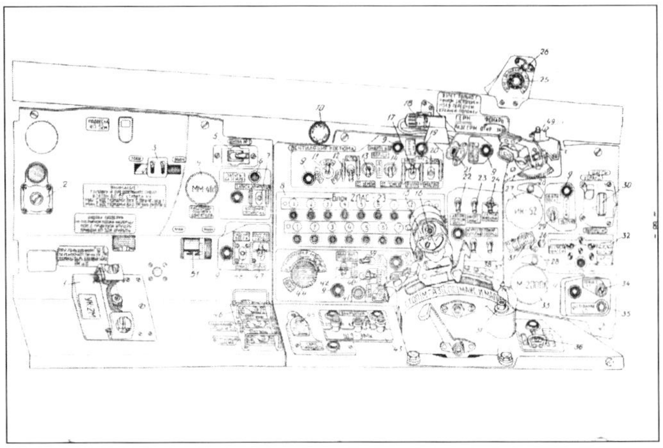

Port side of the cockpit. (Technical Manual)

ASP sight.

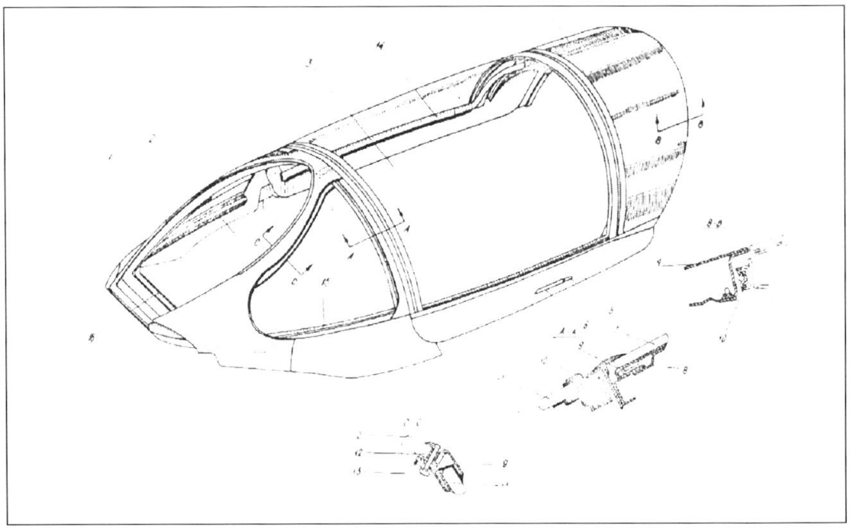

Canopy scheme. (Technical Manual)

The KM-1 ejection seat. (Technical Manual)

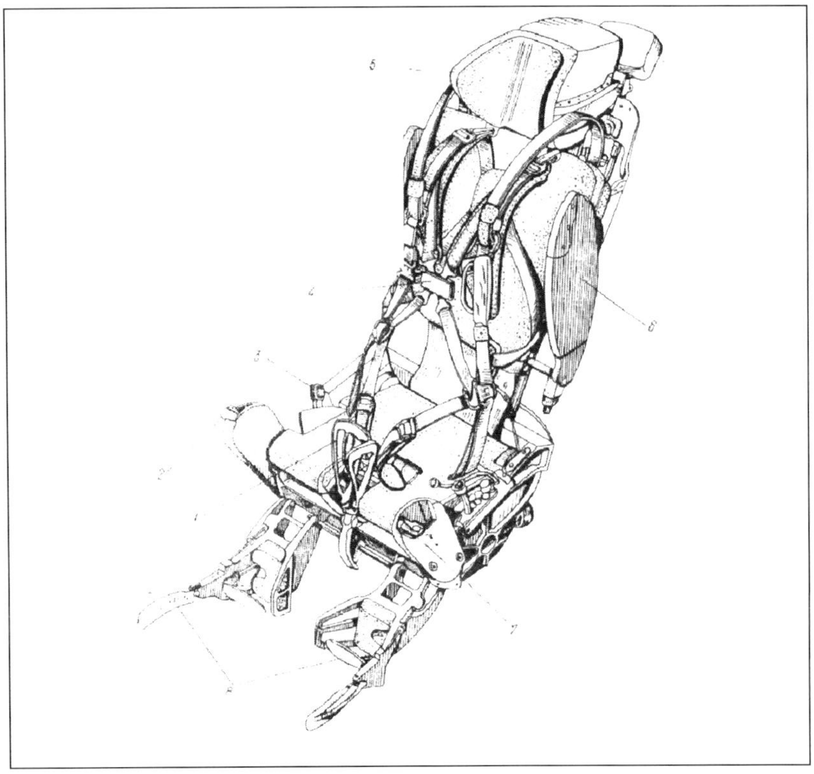

The KM-1 ejection seat.

Cockpit canopy, hinged to starboard. Rear view periscope mirror is visible. (Dariusz Karnas archive)

Undercarriage

Nose wheel leg. (Dariusz Karnas archive)

Nose wheel fork and hydraulic steering unit. (Dariusz Karnas archive)

Main wheel leg. Main wheel doors and wheel bay details. (Dariusz Karnas)

Main wheel leg and wing pylon with launch rail. (Dariusz Karnas)

Main wheel. (Dariusz Karnas)

Armament

R-3S IR missile on outboard pylon. (Leszek Jóźwik)

GSh-23 cannon ammunition loading hatch. (Dariusz Karnas archive)

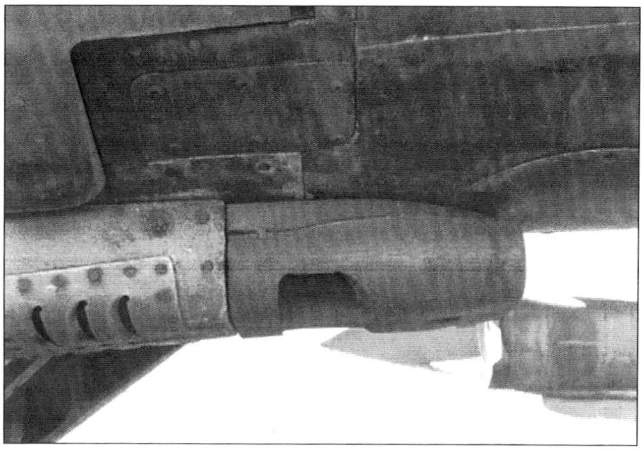

GSh-23 cannon details. (Dariusz Karnas)

Air to air RS-2 US missile.
(Dariusz Karnas archive)

Rockets pods. UB-32 on inboard pylon and UB-16 on outboard pylon.
(Dariusz Karnas)

R-60 dummy missile.
(Dariusz Karnas)

Cockpits details. (Dariusz Karnas archive)

***Below:** UB-16 rockets pod on inboard pylon. Central external fuel tank 490 liters capacity is visible. (Dariusz Karnas)*

Fin spar and ribs structure. RAU-107 autopilot servo, emergency hydraulic pump and flight recorder SARPP-12G (orange colour box). (Dariusz Karnas)

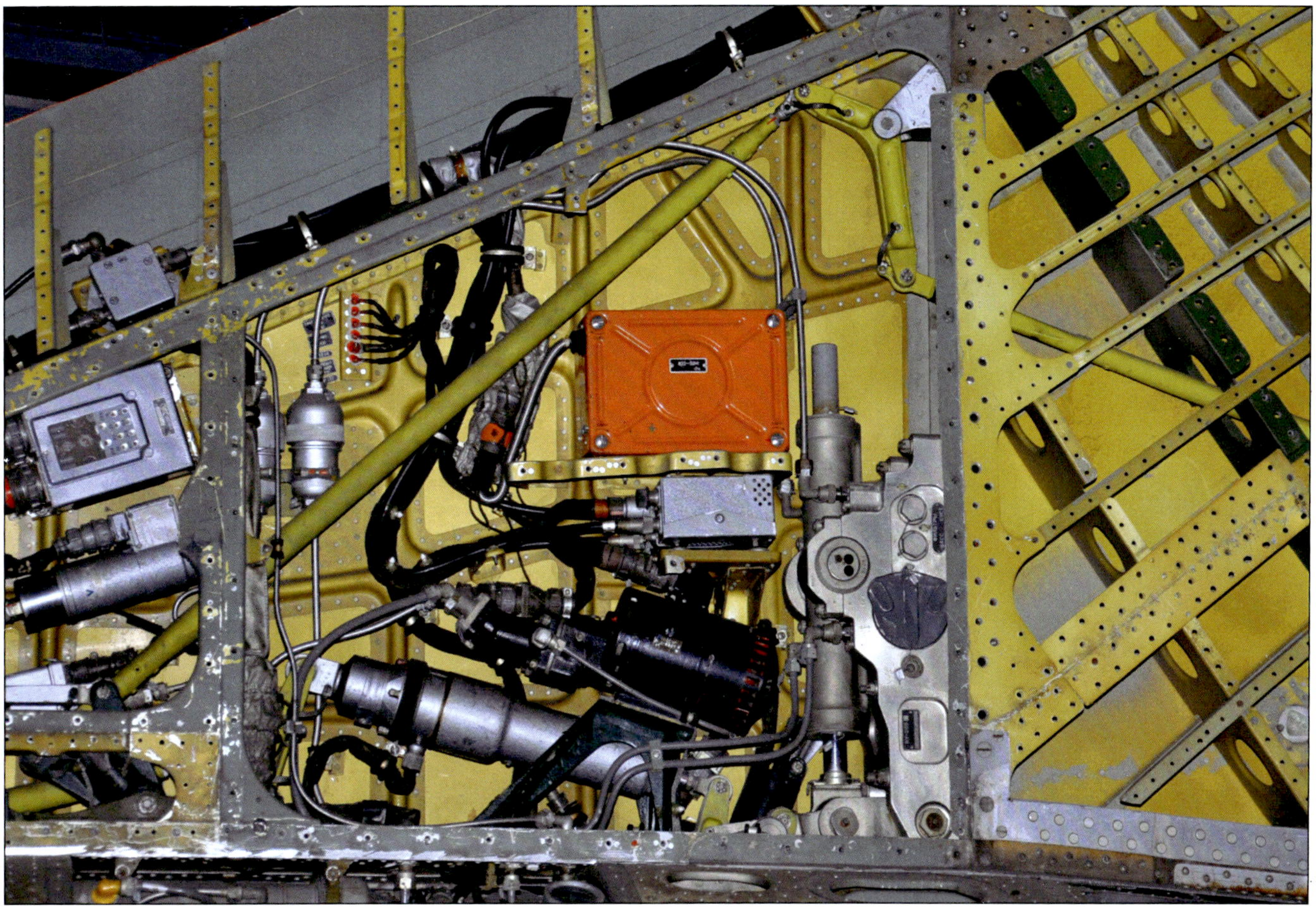

Fin spar and ribs structure. RAU-107 autopilot servo, emergency hydraulic pump, flight recorder SARPP-12G (orange colour box) and fuel ducts. (Dariusz Karnas)

Port main gear bay. (Dariusz Karnas)

Various bombs mounted on under wing pylons. (Dariusz Karnas)

MiG-21MF "6715" (s/n 96A6715), 10. elt. Łask air base, August 2002, pilot – Capt. Jacek Laszczyk. The aircraft wears three-tone gray camouflage. The underside is painted light gray (FS 36495), while the upper side is painted light gray (FS 36320) and dark gray (FS 36118, a mix of three colors) disruptive pattern. The old badge of the 10. elt is painted on the nose and new Pattern 1986 "Dragon" squadron is painted on the vertical stabilizer.

FS 36118 FS 36320 FS 36495

Artur Juszczak

MiG-21MF "6715" (s/n 96A6715), 10. elt. Łask air base, August 2002, pilot – Capt. Jacek Laszczyk.